Guide to Stereochemistry:

A Detailed Guide

Compilation Authored By -

Bishal Baishya

DISCLAIMER:

Henceforth in intentionally or unintentionally anyone else's work has been involved then the person's contribution shall be mentioned in the subsequent revised editions. But there shall be no royalty meant, (unless rare cases as per situation demands) as this book shall be available for informative purposes to all students of College or University Level. Any error or mistake is highly regretted. Informing those errors are welcome at:
contact@bishalbaishya.tk

ACKNOWLEDGMENTS

This is to acknowledge that the information present in this book has been compiled from different sources, by ideas and knowledge. By no means intentionally or unintentionally there has been a direct copying of Information. In some cases the data has also been used from Wikipedia Sources for better reference/understanding.

The information taken from Wiki sources was available under CC3.0 license and henceforth the author if cited as per data available there.

Dedication

This book is dedicated to my Parents and my Teachers, without whom this work was impossible.

Table of Contents

Preface

PREFACE

This book has been written with utmost assurance that this book fulfils the needs of all High-School, College and University students. The book has been writer with utmost care that this book is easy to read and clearly understand the entire complex issues behind the topic "Stereochemistry". The Book has been written after confirming various sources and the information provided are well verified so as to produce simplified but strong Foundational & Complex Knowledge in Students. Still the feedback and corrections are welcome.

– Author

Description of Term:

Stereochemistry, a sub-discipline of chemistry, involves the study of the relative spatial arrangement of atoms that form the structure of molecules and their manipulation. An important branch of stereochemistry is the study of chiral molecules.

Stereochemistry is also known as 3D chemistry because the prefix "stereo-" means "three-dimensionality".

The study of stereochemistry focuses on stereoisomers and spans the entire spectrum of organic, inorganic, biological, physical and especially supramolecular chemistry. Stereochemistry includes methods for determining and describing these relationships; the effect on the physical or biological properties these relationships impart upon the molecules in question, and the manner in which these relationships influence the reactivity of the molecules in question (dynamic stereochemistry).

History of the Term:

Louis Pasteur could rightly be described as the first Stereo-chemist, having observed in 1849 that salts of tartaric acid collected from wine production vessels could rotate plane polarized light, but that salts from other sources did not. This property, the only physical property in which the two types of tartrate salts differed, is due to optical isomerism. In 1874, *Jacobus Henricus van't Hoff* and *Joseph Le Bel* explained optical activity in terms of the tetrahedral arrangement of the atoms bound to carbon.

Drawing of organic molecules:

In order to learn and communicate organic chemistry you must learn the language of organic chemistry. Pictures play an important role in this language, particularly pictures that convey the 3-dimensional properties of an object. You will have to practice drawing on your own but the primer below gives you some simple conventions (tricks) that chemists use to make pictorial communication easier.

Line Segment Drawings: *Freshman Chemistry* presents chemical formulae to us in the following manner:

$$
\begin{array}{ccccc}
 & H & & H & & H \\
 & | & & | & & | \\
H- & C & - & C & - & O \\
 & | & & | & & \\
 & H & & H & &
\end{array}
$$

This is useful for learning the basics and for expressing the details of a small structure, but can be quite cumbersome for expressing ideas about larger molecular structures. For example, this steroidal fragment shown in freshman format is almost unintelligible; it also requires a lot of time to draw.

Many of the structures of importance to organic chemistry are large and as such the old freshman format is no longer useful. Fortunately, organic chemistry deals with compounds containing a high percentage of carbon and hydrogen, and the valences of the two, 4 and 1 respectively, remain constant for most compounds. Therefore, we can use a different drawing convention, one where the carbons are represented as the vertices at the end of each line segment (bond), and the hydrogens are omitted from the drawing but are assumed to be there in the number necessary to give carbon its full valence of four. Take for example hexane. On the left is the freshman format and on the right is the line-segment format.

In all our discussion, we will use the line segment format most of the time and add other drawings to emphasize details. Here again is the structure of the steroidal fragment from the first example but now shown in the line-segment format.

Some important concepts in stereochemistry:

- **Elements of symmetry:**

Symmetry plays a central role in the analysis of the structure, bonding, and spectroscopy of molecules. The symmetry of a molecule is determined by the existence of **symmetry operations** performed with respect to **symmetry elements**. A symmetry element is a line, a plane or a point in or through an object, about which a rotation or reflection leaves the object in an orientation indistinguishable from the original. Four fundamental elements of symmetry are encountered in organic molecules. The various symmetry elements and operations are described in the table below.

Symmetry Elements and Operations

Element	Operation	Symbol
Symmetry plane	reflection through plane	σ
Inversion center	inversion: every point x,y,z translated to $-x,-y,-z$	i
Proper axis	rotation about axis by $360/n$ degrees	C_n
Improper axis	1. Rotation by $360/n$ degrees 2. reflection through plane perpendicular to rotation axis	S_n

- S_2 axis is equivalent to i and S_1 axis is equivalent to σ.

- All molecules in the universe have C_1 axis. It is known as trivial axis.

- **Dissymmetric molecules** do not possess σ, i and S_n symmetry elements but it may or may not contain C_n axis.

- **Asymmetric molecules** do not possess any of these symmetry elements other than C_1 axis. Thus, asymmetry is a special case of dissymmetry.

- All asymmetric molecules are dissymmetric molecules but all dissymmetric molecules are not asymmetric molecules.

- The ability of a molecule to rotate the plane of a plane polarised monochromatic light is known as optical activity. Such molecules are known as **optically active** or **chiral** compounds.

- All asymmetric and dissymmetric molecules are optically active. Thus, for a molecule to be optically active, it must not contain σ and i.

- Optically inactive compounds contain σ and i.

Relationship between objects and their mirror images

Symmetric molecules are superimposable with their mirror images. They are one and the same. Asymmetric molecules are non-superimposable with their mirror images.

SOME DEFINITIONS

Constitutional Isomers: Isomers which differ in "connectivity". The latter term means that the difference is in the sequence in which atoms are attached to one another. These are also called Conformer.

Many definitions that describe a specific conformer (from IUPAC Gold Book) exist, developed by *William Klyne* and *Vladimir Prelog*, constituting their **Klyne–Prelog system** of nomenclature:

- a torsion angle of ±60° is called **gauche**
- a torsion angle between 0° and ± 90° is called **syn (s)**
- a torsion angle between ± 90° and 180° is called **anti (a)**
- a torsion angle between 30° and 150° or between −30° and −150° is called **clinal**
- a torsion angle between 0° and 30° or 150° and 180° is called **periplanar (p)**
- a torsion angle between 0° to 30° is called **synperiplanar** or **syn-** or **cis-conformation (sp)**

- a torsion angle between 30° to 90° and −30° to −90° is called **synclinal** or **skew (sc)**
- a torsion angle between 90° to 150°, and −90° to −150° is called **anticlinal (ac)**
- A torsion angle between ± 150° to 180° is called **antiperiplanar (ap).**

Torsional strain results from resistance to twisting about a bond.

Stereoisomers - Compounds that have the same molecular formula and the same connectivity of atoms, but different relative arrangement of the atoms in 3-dimensional space. For

example, trans-but-2-ene and trans-but-2-ene are stereoisomers.

Enantiomers – A pair of stereoisomers of a molecule having non-superimposable mirror image relationship. Molecules that form a pair of enantiomers are chiral. A chiral molecule is related to its own mirror image in the way that your left hand is related to your right hand.

All chiral molecules have at least one of the three kinds of chiral centers, the centre of chirality, the axis of chirality or the plane of chirality. Chiral molecules are optically active compounds. Molecules without any chiral centre are achiral and optically inactive.

CHIRAL AND ACHIRAL MOLECULES

2-BUTANOL (CHIRAL)

Mirror Plane

ORIGINAL MOLECULE

MIRROR IMAGE ISOMER

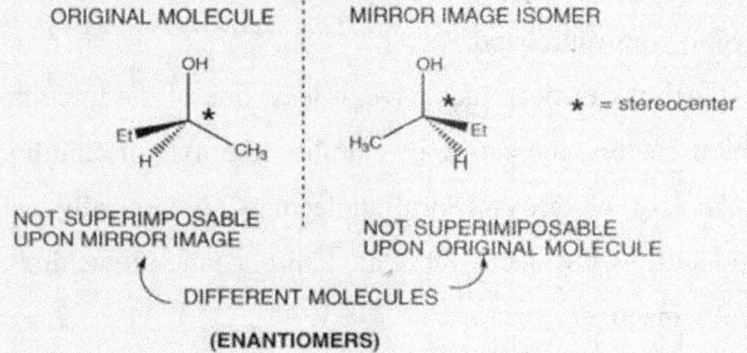

★ = stereocenter

NOT SUPERIMPOSABLE
UPON MIRROR IMAGE

NOT SUPERIMPOSABLE
UPON ORIGINAL MOLECULE

DIFFERENT MOLECULES

(ENANTIOMERS)

2-PROPANOL (ACHIRAL)

MIRROR PLANE

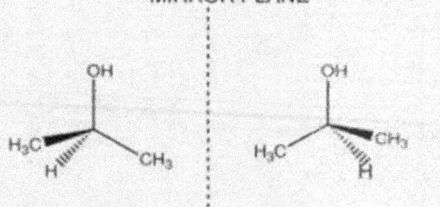

NO STEREOCENTER.
C2 HAS ONLY **THREE**
NON-EQUIVALENT
GROUPS.

THE ORGINAL MOLECULE AND THE MIRROR IMAGE
ARE **SUPERIMPOSABLE** AND THEREFORE REPRESENT
MOLECULES OF THE SAME SUBSTANCE. THE
MOLECULES ARE **ACHIRAL** AND DO NOT HAVE
ENANTIOMERS.

Enantiomers are identical in most physical and chemical properties, but they are different in the way they react with other chiral molecules and in the way they interact with polarized light. Their interaction with polarized light is called optical activity.

Chiral or asymmetric carbon: A tetrahedral carbon atom bearing four different substituents.

A methane derivative, a sp^3 hybridized carbon atom with four different ligands, is a good example of compounds with a centre of chirality. Such compounds can be represented as CXYZW. This type of carbon atom is called an *asymmetric carbon atom*. An asymmetric carbon atom is sometimes indicated by an asterisk (*) in the chemical formula.

Relationships between chiral centers and chiral molecules:
The term chiral centre refers to an atom in the molecular structure. The term chiral molecule refers to the entire molecule. The presence of one chiral centre renders the entire molecule chiral.

The presence of two or more chiral centers may or may not result in the molecule being chiral.

In the examples given below the chiral centers are indicated with an asterisk. The vertical broken line represents a plane of symmetry.

Ibuprofen. One chiral center renders the molecule chiral

cis-1,2-dimethylcyclohexane is an achiral molecule

trans-1,2-dimethylcyclohexane is a chiral molecule

Meso compounds- A Meso compound or Meso isomer is an optically inactive member of a set of stereoisomers, at least two of which are optically active. This means that despite containing two or more chiral centers it is not chiral. A Meso compound is "superimposable" on its mirror image.

In general, a Meso compound should contain two or more identical substituted stereocenters. Also, it has an internal symmetry plane that divides the compound in half. These two halves reflect each other by the internal mirror. The stereochemistry of stereocenters should "cancel out". What it means here is that when we have an internal plane that splits the compound into two symmetrical sides, the stereochemistry of both left and right side should be opposite to each other, and therefore, result in **optically inactive**. Cyclic compounds may also be Meso.

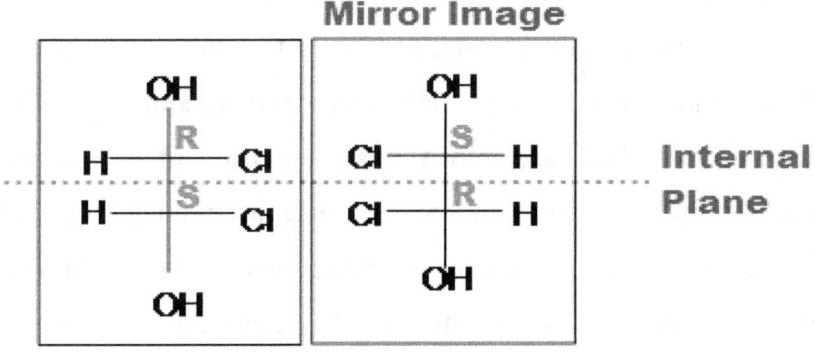

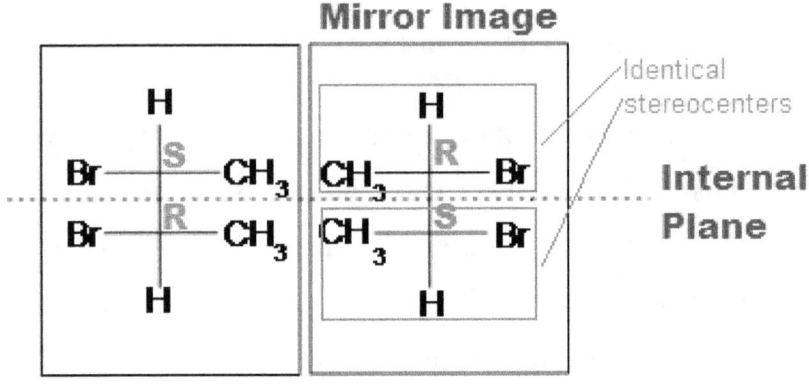

Mirror Image

Identical stereocenters

Internal Plane

Diasteromers - Stereoisomers of a molecule which are not enantiomers (or mirror images to each other).). Diastereomers are stereoisomers that are not mirror images of one another and are non-superimposable on one another. Stereoisomers with two or more stereocenters can be diastereomers.

Diastereomers are stereoisomers that are not related as object and mirror image and are not enantiomers. Unlike enantiomers which are mirror images of each other and non-superimposable, diastereomers are not mirror images of each other and non-superimposable. Diastereomers can have different physical properties and reactivity. They have different melting points and boiling points and different densities. They have two or more stereocenters.

It is easy to do mistake between diastereomers and enantiomers. For example, we have four stereoisomers of 3-bromo-2-butanol. The four possible combinations are SS, RR, SR and RS (see below). One of the molecules is the enantiomer of its mirror image molecule and diastereomers of each of the other two molecules (SS is enantiomer of RR and diastereomers of RS and SR). SS's mirror image is RR and they are not superimposable, so they are enantiomers. RS and SR are not mirror image of SS and are not superimposable to each other, so they are diastereomers.

(2S, 3S) -3-Bromo-2-Butanol
Mirror Plane
(2R, 3R) -3-Bromo-2-Butanol

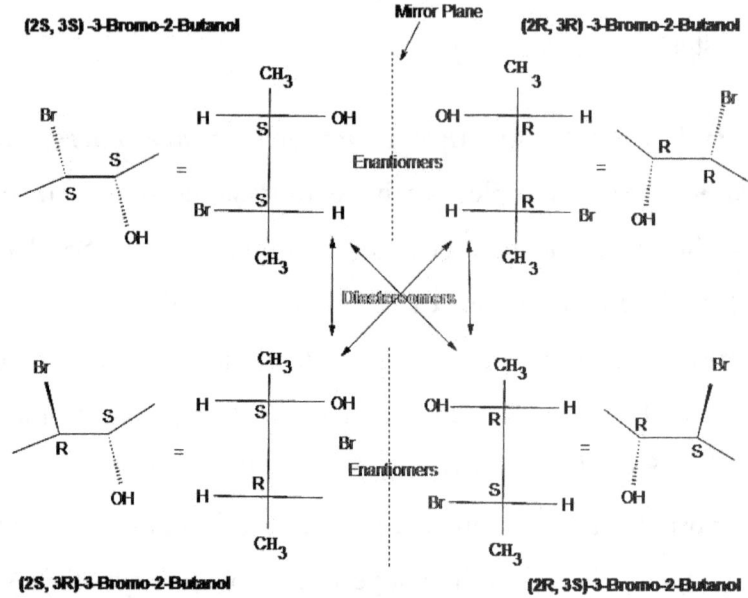

Enantiomers

Diastereomers

Enantiomers

(2S, 3R)-3-Bromo-2-Butanol
(2R, 3S)-3-Bromo-2-Butanol

Optical activity:

The ability of a molecule to rotate the plane of a plane polarised monochromatic light. Such molecules are known as **optically active** compounds.

Light is polarized when it passes through a Nicol prism or Polaroid. The transmitted light is in a single oscillating plane. the plane-**polarized light** (polarized by the first Polaroid) passes through the second polaroid when placed so that its lines are those in the first one, but the polarized light is completely blocked when lines of the second polaroid are perpendicular to those in the first one.

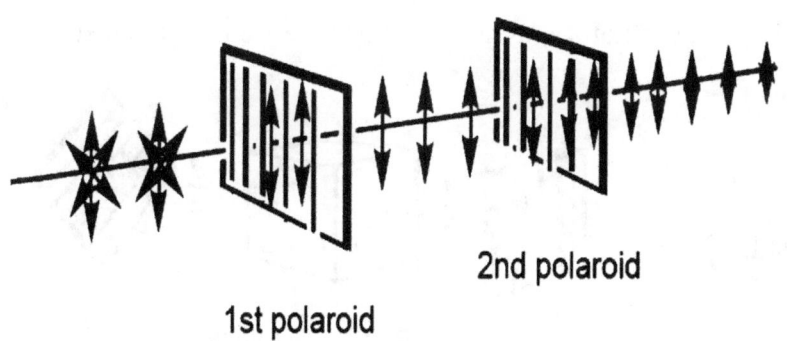

2nd polaroid

1st polaroid

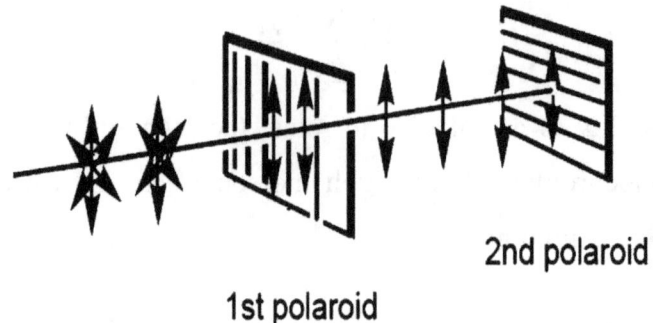

2nd polaroid

1st polaroid

Compounds that can rotate the plane of plane-polarized light are **optical active**. Optically active compounds exist as a pair of **enantiomers**. If one of the enantiomers rotates the polarized light plane clockwise, the other enantiomer rotates it the same amount counter clockwise to an equal degree. Clockwise rotation is "**dextrorotatory**", and counter clockwise rotation is "**laevorotatory**". For dextrorotatory, symbols (+) or *d*, and for laevorotatory, symbols (-) or *l* are frequently used.

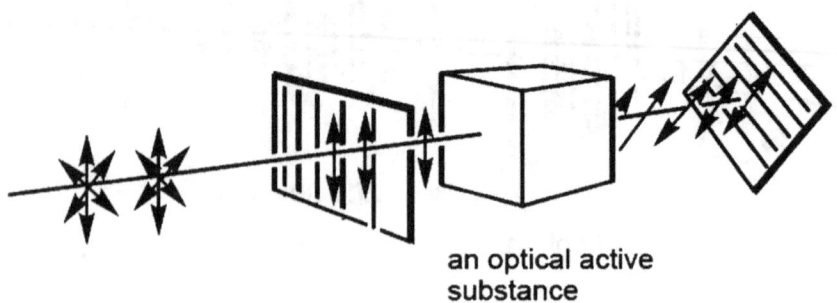

an optical active substance

The amount of rotation depends not only on the type of the discussed compound, but also on its concentration in the solution, the length of the light path through the solution, the solvent, the wavelength of the light employed, and the temperature. The specific rotation [α] is defined for a given compound as the rotation (°) induced by 1 g/ml solution, 10 cm long, under a specified condition (the temperature, wavelength of the light used, the solvent) as shown below.

$$[\alpha]_{\lambda}^{t} = \frac{\text{observed rotation (°)}}{\text{length of the sample(dm) x concentration(g/ml)}}$$

Where λ is the wavelength of the light and **t** is the temperature.

Dextrorotatory - Ability of chiral substances to rotate the plane of polarized light to the right (see above-already discussed).

Laevorotatory - Ability of chiral substances to rotate the plane of polarized light to the left (see above-already discussed).

Specific rotation - The measured angle of rotation of polarized light by a pure chiral sample under specified standard conditions.

Racemic mixture, racemic modification, or racemate- A mixture consisting of equal amounts of enantiomers.

The two compounds forming a pair of enantiomers have the same physical (*e.g.*, melting point) and chemical (*e.g.*, acidity) properties except for optical rotation and reactivity to chiral reagents. A 50:50 mixture of enantiomers is called the **racemate** and designated as (±) or *dl*. Racemates are optically inactive because the effect of one enantiomer is cancelled by the effect of the other enantiomer. The chemical property of the solution of a racemate and that of a single enantiomer are identical, though sometimes they are different in the solid state. Separation of racemates into their component enantiomers is called **optical resolution** or simply **resolution**.

There are three methods of resolution.

1) <u>Physical resolution</u>: manual separation when enantiomers crystallize in different crystal forms.

2) <u>Chemical resolution</u>: separation by chemical methods, separating the diastereomers obtained by a reaction with chiral reagents.

3) <u>Biological resolution</u>: use of a specific enzyme which selectively consumes one of the enantiomers.

Optical purity:

The difference in percent between two enantiomers present in a mixture in unequal amounts. For example, if a mixture contains 75% of one enantiomer and 25% of the other, the optical purity is 75-25 = 50%.

Thus, if the specific rotation of a pure single enantiomer is known, it is easy to determine the purity of a sample containing both enantiomers in unequal amounts. The % OPTICAL PURITY = specific rotation of the sample/specific rotation of the pure enantiomer. This particular measure of optical purity is often called enantiomeric excess (or ee) because it gives %R - %S. A small problem (admittedly very small, mathematically) arises in converted the ee (enantiomeric excess) into a specific composition given in terms of %R and %S. One simple way of doing this is as follows: If the enantiomeric excess of the R enantiomer is, for example, 80%, this means that there is 80% of the R enantiomer plus 20% of the racemic mixture (not 20%S). Since the racemic mixture is 10%R and 10%S, the composition of the mixture is 90% R and 10%S. Remember: ee represents not the % of one of the enantiomers, but the difference between the % of one pure enantiomer and the % of racemic mixture).

Molecules with chirality centers cause the rotation of plane polarised light and are said to be "optical active" (hence the term optical isomers). Enantiomeric molecules rotate the plane in opposite directions but with the same magnitude.

This provides a means of measuring the **"optical purity"** or **"enantiomeric excess** (ee)" of a sample of a mixture of enantiomers. These two terms mean the same thing: How much more of one enantiomer is there than the other? Specific rotation is a physical property like boiling point and can be looked up in references. It is defined according to the following equation based on the experimental measurements:

$$\textbf{Specific rotation } [\alpha]_D = \alpha_{obs} / c\,l$$

Where "α_{obs}" is the experimentally observed rotation, "c" is the concentration in g/ml and "l" is the path length of the cell used expressed in dm (10 cm).

The most important factor is that two enantiomers will have the **same magnitude** specific rotations but in *opposite directions.*

For example: (S)-bromobutane has a specific rotation of +23.1°, therefore, (R)-Bromobutane has a specific rotation of -23.1°

As a consequence of this, a 50:50 mixture of the two enantiomers will not rotate plane polarised light because the effects of the two enantiomers cancel each other out, molecule for molecule. This type of mixture is called a *racemate* or a *racemic mixture*. The specific rotation of a racemic mixture is zero.

The optical purity of a mixture of enantiomers is given by:

% Optical purity of sample = 100 * (specific rotation of sample) /

(Specific rotation of a pure enantiomer)

Based on the above example data for the bromobutanes:

Optical purity of a racemic mixture = 100 * (0°) / (+23.1°) = 0% *i.e. there is no one enantiomer present in excess.*

Another way to express optical purity is as the "**Enantiomeric excess**" or "**ee**":

Enantiomeric excess % = 100 * ([*d*] - [*l*]) / ([*d*] + [*l*])

Where [ ] indicates the concentration of *d* or *l* enantiomer. Repeating the above example (to show the identity) in a racemic mixture [d] = [l], therefore,

$$\textbf{ee\%} = 100 * 0 / ([d] + [l]) = 0\%$$

Or, if we use the % as concentrations, then [d] = [l] = 50 %

We get,

$$\textbf{ee\%} = 100 * (50 - 50) / (50 + 50) = 0 \%$$

So, what about for a pure enantiomer, say 100% *d* ?

Optical purity % = 100 * (-23.1° / -23.1°) = 100 %,

Or, ee% = 100 * (100% - 0%) / (100% + 0%) = 100%

So, finally, what about another mixture of the 2-bromobutanes of measured specific rotation = +9.2°? Well the fact that the sign is positive tells us that in this case the *d* enantiomer is the dominant one.

Optical purity % = 100 * (+9.2° /+ 23.1°) = 40 % , *i.e. there is a 40% excess of d over l.*

This corresponds to a mixture of 70% d and 30% l. How do you get this quickly?

Well, if there is a 40% excess of *d*, then the 60% leftover must be equal amounts of both *d* and *l i.*e. 30% of each. So the total amount of *d* isomer is 30% + 40% excess = 70%.

Absolute configuration:- A description of the precise 3-dimensional topography of the molecule. In other words, the exact spatial orientations of the substituents of the isomers (cis-trans or R-S)

Relative configuration:- A description of the 3-dimensional topography of the molecule relative to an arbitrary standard. Absolute and relative configurations may or may not coincide

Flying-wedge formula: Often you must convey the three dimension nature of a structure. One simple way to do this is through the bond-wedge convention. This is called a convention because you can't actually show 3-dimensions on a flat piece of paper so we must agree to some convention of what certain shapes are going to tell us. In this convention, a solid wedge means that the bond is coming out of the plane of the paper toward us; a dashed wedge means that the bond is going into the plane of the paper away from us; finally a simple line means that the bond is in the plane of the paper.

bonds in the plane of the paper

bond bellow the plane of the paper

bond above the plane of the paper

(R) - Lactic acid

The Fischer Projection:

The flying-wedge representations of stereochemistry can often become cumbersome, especially for large molecules which contain a number of stereocenters. An alternative way to represent stereochemistry is the **Fischer Projection**, which was first used by the German chemist Emil Fischer. The Fischer projection is a two-dimensional representation of a three-dimensional organic molecule by projection. According to the rule of the Fischer projection, the asymmetric carbon atom lies in the plane of the paper, the horizontal bonds project above the paper and the vertical bonds beneath. The Fischer Projection represents every stereocenter as a cross. In other words, the Fischer projections **7** and **8** are equivalent to the flying-wedge drawings **1** and **2**, respectively.

Flying-wedge
drawing

Tetrahedral
drawing

Fischer
projection

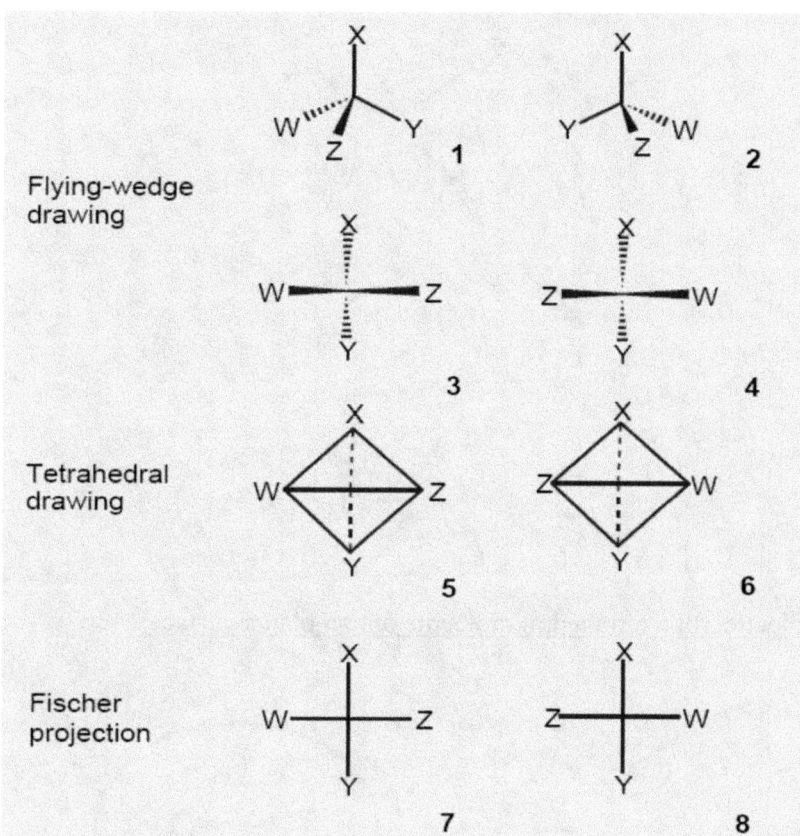

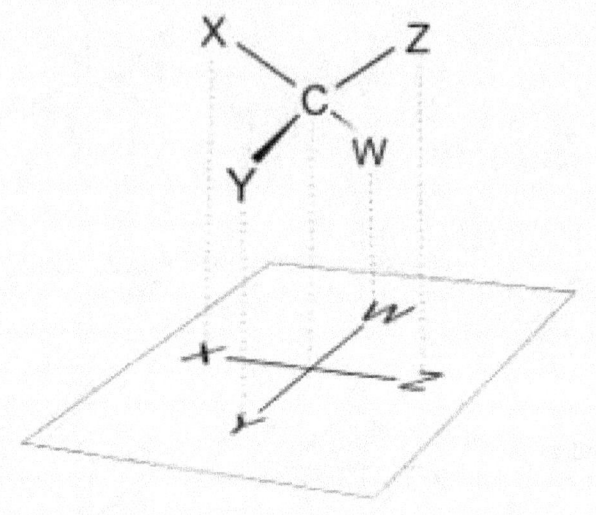

Projection of a tetrahedral molecule onto a planar surface:

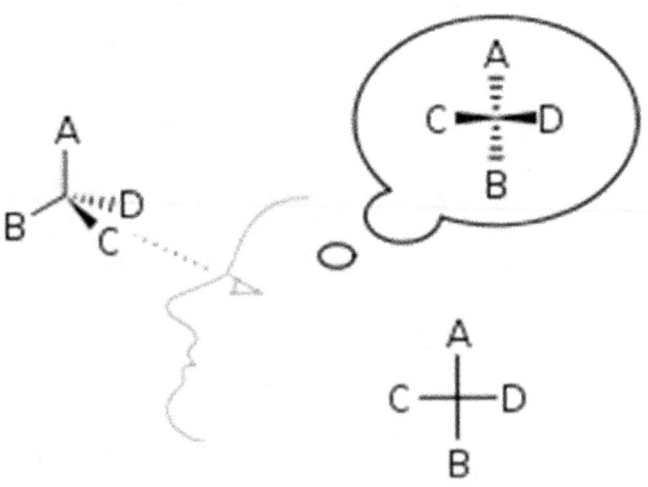

Visualizing a Fischer projection:

When working with Fischer Projections, keep in mind the following rules:

1. Because the "up" and "down" aspects of the bonds don't change, a Fischer projection may be rotated by 180 degrees without changing its meaning.

2. A Fischer projection cannot be rotated by 90 degrees. Such a rotation typically changes the configuration to the enantiomer.

3. To find the enantiomer of a molecule drawn as a Fischer projection, simply exchange the right and left horizontal bonds.

4. To determine whether the molecule in Fischer projection is a Meso compound, draw a horizontal line through the centre of the molecule and determine whether the molecule is symmetric about that line.

1. Rotation by 180° is allowed

same

2. Rotation by 90° is not allowed

enantiomer

3. Enantiomer results from swapping left and right bonds

enantiomer

4. Meso compounds can be found by examining symmetry about horizontal plane

meso (achiral)   chiral

Absolute configuration—*R, S*-Convention:

Since two enantiomers are different compounds, we will need to have nomenclature which distinguishes them from each other. The convention which is used is called the (R,S) system because one enantiomer is assigned as the R enantiomer and the other as the S enantiomer. What are the rules which govern which are which??

The practice of R, S nomenclature is as follows:

For each chiral centre (compounds with many centers should also be considered),

(1) Make sure that the molecule contains at least one chiral centre. The fact that a 3-dimensional formula of a molecule is given does not imply that there is(are) chiral centre(s).

(2) The priorities of four substituents around a chiral centre are decided by assigning **CIP rule-based** priorities to the atoms which are directly attached to the chiral centre.

(3) An alphabet is assigned to each ligand according to the decreasing order of priority; a > b> c > d. Some books use 1 > 2> 3 > 4 instead of a > b> c > d.

(4) Suppose you are looking down the bond from the asymmetric carbon atom toward the ligand of lowest priority (d). The other three ligands (a, b and c) will be facing you. Connect these three ligands with an arrow running from highest to lowest priority (a> b > c).

It is important to remember that there is no simple or obvious relationship between the R or S designation of a molecular configuration and the experimentally measured specific rotation of the compound it represents. *In order to determine the true or "absolute" configuration of an enantiomer, it is necessary either to relate the compound to a known reference structure, or to conduct a rather complex X-ray analysis on a single crystal of the sample.*

CIP RULES:

The Cahn–Ingold–Prelog priority rules, CIP system or CIP conventions (after Robert Sidney Cahn, Christopher Kelk Ingold and Vladimir Prelog) are a set of rules used in organic chemistry to name the stereoisomers of a molecule. A molecule may contain any number of stereocenters and any number of double bonds, and each gives rise to two possible configurations. The purpose of the CIP system is to assign an R or S descriptor to each stereocenter and an E or Z descriptor to each double bond so that the configuration of the entire molecule can be specified uniquely by including the descriptors in its systematic name.

Assignment of priorities: CIP rules for R and S nomenclatures are as follows:

1. Compare the atomic number (Z) of the atoms directly attached to the stereocenter; the group having the atom of higher atomic number receives higher priority. In case there are isotopes, use the mass number instead, since they have the same atomic number.

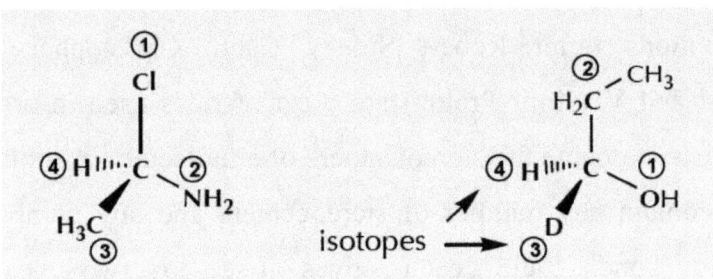

2. If two or more of the atoms directly attached to the chiral centre are of the same type, look at the next atom to break the tie (*do not do this unless there is a tie*). Repeat this process until the tie is broken. It is important to emphasize that in trying to break ties, one looks at the atoms **directly attached to the element under observation before looking at any others**. Study the following examples very carefully to make sure this point is clear.

3. For the purpose of assigning priority, double and triple bonds are spilt into two and three single bonds respectively by replication of the atom at the other end of the multiple bonds. Replicated atoms are enclosed in parentheses in the expanded form of the group. Each replicated atom except for H atom is converted to single bond tetra-covalence by adding so called phantom atoms which are denoted by cipher zero and shown as subscript. Phantom atoms are imaginary having atomic number zero. In summary, we can say:

The coordination number of non-hydrogen atoms is assumed to be 4, i.e. atoms bonded with multiple bonds are considered to be bonded to multiple atoms, e.g. carbonyl carbon is treated as if it was bonded to two oxygen atoms, and carboxyl carbon as if it was bonded to three oxygens (these are then called phantom atoms).

How four common groups are treated in the Cahn-Ingold-Prelog			
GROUP	**Treated as if it were**	**GROUP**	**Treated as if it were**
$-\overset{\overset{H}{\mid}}{C}=O$	$-\overset{\overset{H}{\mid}}{\underset{\underset{O_{000}}{\mid}}{C}}-O_{00}-C_{000}$	$-CH=CH_2$	$-\overset{\overset{H}{\mid}}{\underset{\underset{C_{000}H}{\mid}}{C}}-\overset{\overset{H}{\mid}}{C}-C_{000}$
$-C\equiv CH$	$-\overset{\overset{C_{000}H}{\mid}}{\underset{\underset{C_{000}C_{000}}{\mid}}{C}}-\overset{}{C}-C_{000}$	$-C_6H_5$	$H-\overset{\overset{C_{000}}{\mid}}{C}-\overset{}{C}-$

A list of common groups of increasing priority according to the above sequence rule is given below:

-I > -Br > -Cl > -SO$_2$R > -SOR > -SR > -SH > -F > -OCOR > -OR > -OH > -NO$_2$ > -NO > -NHCOR > -NR$_2$ > -NHR > - NH$_2$ > -CX$_3$ > -COX > -CO$_2$R > -CO$_2$H > -CONH$_2$ > -COR > -CHO > -CR$_2$OH > -CH(OH)R > -CH$_2$OH > -C≡CR > -C$_6$H$_5$ (phenyl) > -C≡CH > -C(R)=CR$_2$ > -CR$_3$ > -CHR$_2$ > -CH$_2$R > -CH$_3$ > D > H

X = Halogen atoms, R = alkyl groups

4. When two substituents are similar chiral centres but differ only in configuration then the chiral centre with R configuration gets priority over the chiral centre with S configuration.

5. Ligands with R, R or S, S precede R, S or S, R.

6. It needs to be mentioned also that two substituents on an atom may, in rare cases, be geometrical isomers. Consider for example the compound (3Z, 6E)-3, 5, 7-trimethylnona-3, 6-diene. It soon becomes clear that the 5-carbon is chiral because it has four different substituents. Thus it is necessary to introduce the rule that the Z-isomer has higher priority than the E-isomer.

Shortcuts for assigning absolute configuration on paper in a flying wedge projection formula:

According to the Cahn-Ingold-Prelog (CIP) convention, when assigning absolute configuration to a chiral carbon the lowest priority group that's attached to that carbon must be pointing away from an observer who is looking at the carbon in question. On paper, that usually means that if the observer is the person looking at the page, then the lowest priority group is pointing away from the observer, going behind the plane of the paper. In a 3-D formula this is indicated thus:

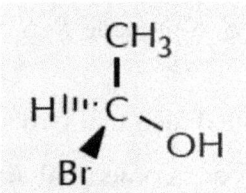

When the formula is given to us in this way, it's easy to assign configuration. All we have to do is assign priorities to the other three substituents and see if they are arranged clockwise or counter clockwise when the observer follows them in order of decreasing priorities. We don't have to mentally reposition either ourselves or the molecule in any way. In the example given above we can see that the central (chiral) carbon has the (S) configuration.

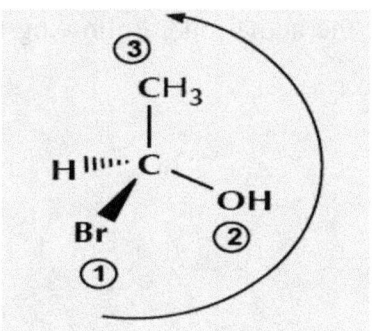

If the lowest priority group is not presented to us already positioned towards the back of the chiral carbon, then it is useful to remember the following basic principle:

"Every time any two substituents are exchanged, the opposite configuration results"

With this in mind, we can encounter two possible scenarios: Either the lowest priority group is positioned in front of the chiral carbon, or on the plane of the paper. If the lowest priority group is positioned in front of the chiral carbon (that is, opposite where it should be, according to the rules) we can still assign configuration by following the arrangement of the other three groups as given to us, but the configuration we obtain will be the opposite of the actual one. Following the same example given above we have:

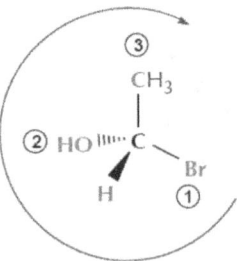

With the lowest priority group positioned in the front rather than towards the back, the central carbon appears to have the (R) configuration. The actual configuration is therefore (S). This is best seen when we rotate the molecule until the H atom is in the correct orientation.

If the lowest priority group is positioned on the plane of the paper, we can momentarily exchange it with whatever group happens to be positioned in the back, then assign configuration, then reverse it.

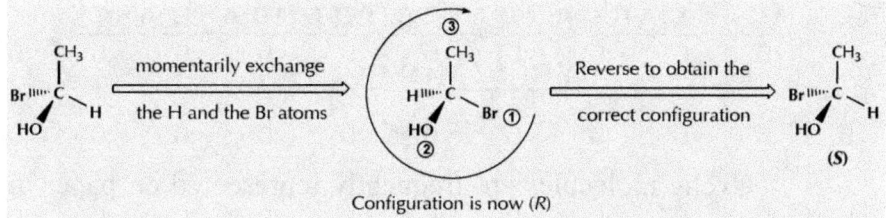

Configuration is now (R)

ASSIGNING ABSOLUTE CONFIGURATIONS IN CYCLIC MOLECULES

Cyclic molecules are frequently represented on paper in such a way that the ring atoms are all lying on the plane of the paper, and substituents are either coming out of the paper towards the front or towards the back. It is therefore easy to assign configuration to any chiral centers forming part of the ring, since the lowest priority substituent will be either pointing to the front or to the back. However, **always make sure there is in fact a chiral centre present**. The fact that a 3-D representation is given does not necessarily mean there is a chiral centre in the molecule.

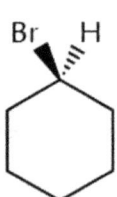

No chiral centers present anywhere

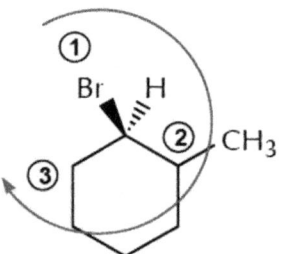

A chiral center is present with the (R) configuration

ASSIGNING ABSOLUTE CONFIGURATIONS IN FISCHER FORMULAS:

The key points to keep in mind regarding Fischer projection formulas are:

1. Horizontal lines represent bonds to the chiral carbon that are coming out of the plane of the paper towards the front, whereas vertical lines represent bonds going behind the plane of the paper towards the back.

2. When the fourth ranked (4) priority group is on the vertical line [either north (top) or south (bottom) position] and the order of priorities follows a clockwise direction, the absolute configuration of the chiral centre will be R.

direction of ① ⟶ ② ⟶ ③	absolute configuration
clockwise	R
counterclockwise	S

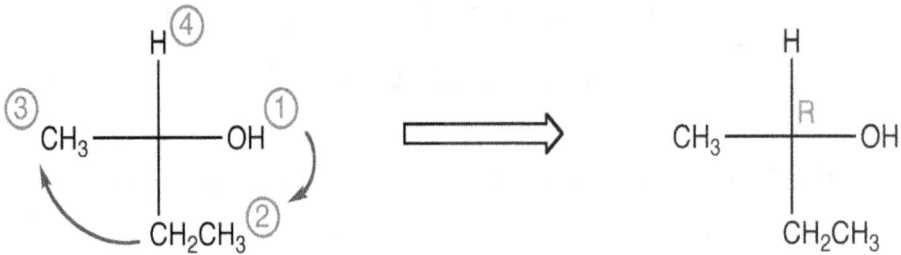

3. When the fourth ranked (4) priority group is on the horizontal line [either right (east) or left (west) position] and the order of priorities follows a clockwise direction, the absolute configuration of the chiral centre will be S.

direction of ① ⟶ ② ⟶ ③	absolute configuration
clockwise	S
counterclockwise	R

Carbon-2 (red dot) has the (R) configuration

Carbon-3 (red dot) also has the (R) configuration

A Fischer projection restricts a three-dimensional molecule into two dimensions. Consequently, there are limitations as to the operations that can be performed on a Fischer projection without changing the absolute configuration at chiral centers. The operations that do not change the absolute configuration at a chiral center in Fischer projections can be summarized as two rules.

Rule 1: Rotation of the Fischer projection by 180° in either direction without lifting it off the plane of the paper does not change the absolute configuration at the chiral center.

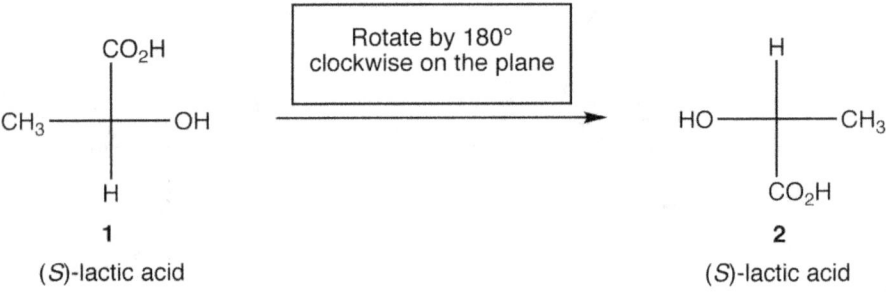

Rule 2: Rotation of three ligands on the chiral centre in either direction, keeping the remaining ligand in place, does not change the absolute configuration at the chiral centre.

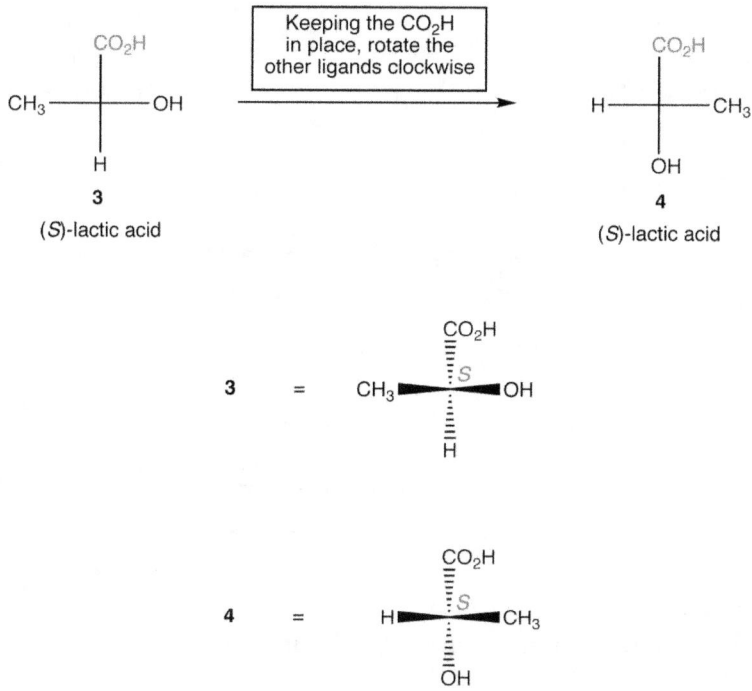

The operations that do change the absolute configuration at a chiral centre in a Fischer projection can be summarized as two rules.

Rule 1: Rotation of the Fischer projection by 90° in either direction changes the absolute configuration at the chiral centre.

Rule 2: Interchanging any two ligands on the chiral centre changes the absolute configuration at the chiral centre.

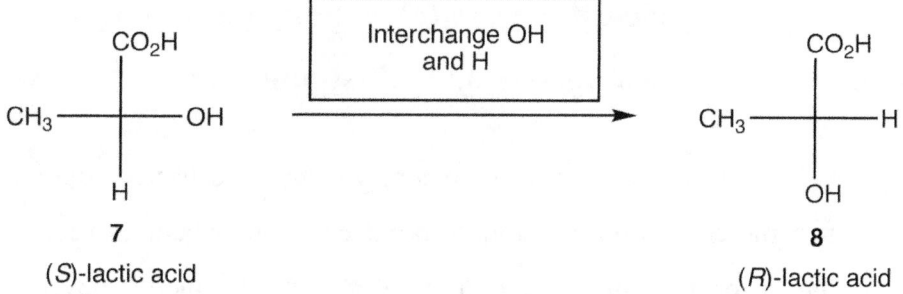

$$7 \quad = \quad CH_3 \overset{CO_2H}{\underset{H}{\overset{\mid}{\longleftarrow S \longrightarrow}}} OH$$

$$8 \quad = \quad CH_3 \overset{CO_2H}{\underset{OH}{\overset{\mid}{\longleftarrow R \longrightarrow}}} H$$

The above rules assume that the Fischer projection under consideration contains only one chiral centre. However, with care, they can be applied to Fischer projections containing any number of chiral centers.

Geometric isomerism (also known as cis-trans isomerism or E-Z isomerism)

These isomers occur where you have restricted rotation in a molecule **about a double bond or a cycloalkane structure due to too high energy requirement.** At an introductory level in organic chemistry, examples usually just involve the carbon-carbon double bond - and that's what this page will concentrate on.

Think about what happens in molecules where there is *unrestricted* rotation about carbon bonds - in other words where the carbon-carbon bonds are all single. The next diagram shows two possible configurations of 1, 2-dichloroethane.

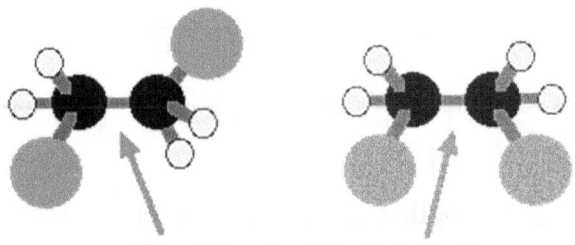

free rotation about this single bond

These two models represent exactly the same molecule. You can get from one to the other just by twisting around the carbon-carbon single bond. These molecules are ***not isomers***.

If you draw a structural formula instead of using models, you have to bear in mind the possibility of this free rotation about single bonds. You must accept that these two structures represent the same molecule.

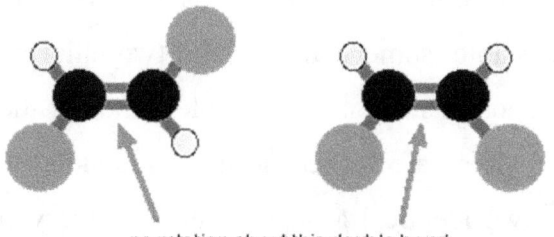

But what happens if you have a carbon-carbon double bond - as in 1, 2-dichloroeth<u>ene</u>?

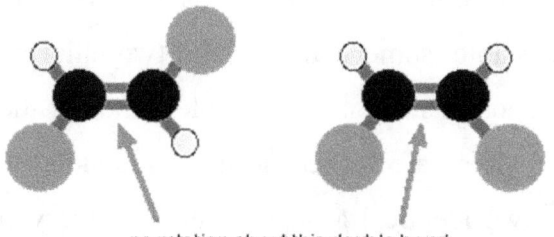

no rotation about this double bond

These two molecules aren't the same. The carbon-carbon double bond won't rotate and so you would have to take the models to pieces in order to convert one structure into the other one. That is a simple test for isomers. If you have to take a model to pieces to convert it into another one, then you've got isomers. If you merely have to twist it a bit, then you haven't!

In the model, the reason that you can't rotate a carbon-carbon double bond is that there are two links joining the carbons together. In reality, the reason is that you would have to break the pi bond. Pi bonds are formed by the sideways overlap between p orbitals. If you tried to rotate the carbon-carbon bond, the p orbitals won't line up any more and so the pi bond is disrupted. This costs energy and only happens if the compound is heated strongly. **Thus, we should not say rotation is impossible; it is the energy barrier that causes the existence of two distinct isomers!**

Drawing structural formulae for the last pair of models gives two possible isomers. In one, the two chlorine atoms are locked on opposite sides of the double bond. In the other, the two chlorine atoms are locked on the same side of the double bond. Thus, two identical groups attached to the vicinal double bonded carbon atoms of the alkene could be positioned on the same side of the alkene or on opposite sides of the alkene. Such compounds are different in chemical and physical properties as well as in their geometry, and are called geometrical isomers. In 2-butene the methyl groups can be located on the same side or on the opposite side of the double bond, giving rise to two geometrical isomers. The isomer with the methyl groups on the same side is called the *cis* isomer (*cis*: from Latin meaning "on this side"), while the isomer with the methyl groups located on opposite sides is called the *trans* isomer (*trans*: from Latin meaning "across" - as in transatlantic).

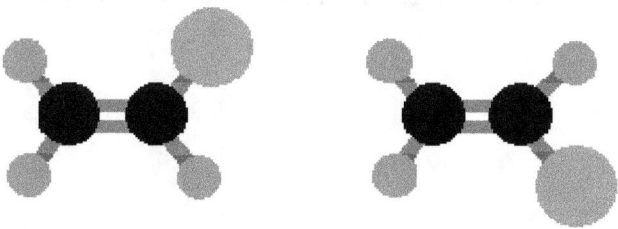

What needs to be attached to the carbon-carbon double bond for an alkene to show geometrical isomerism?

Think about this case:

Although we've swapped the right-hand groups around, these are still the same molecule. To get from one to the other, all you would have to do is to turn the whole model over. ***You won't have geometric isomers if there are two identical groups attached to the same sp² carbon atom*** - in this case, the two pink groups on the left-hand end. So..... There must be two different groups on the left-hand sp² carbon and two different groups on the right-hand one. The cases we've been exploring earlier are like this:

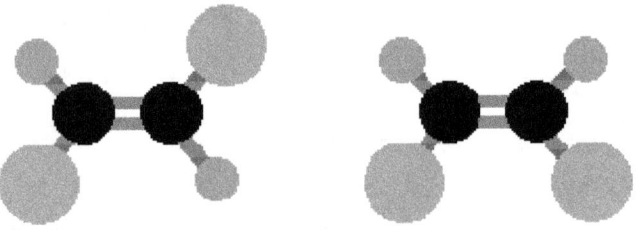

Here the left isomer is *trans* and right one is *cis*. But you could make things even more different and still have geometric isomers:

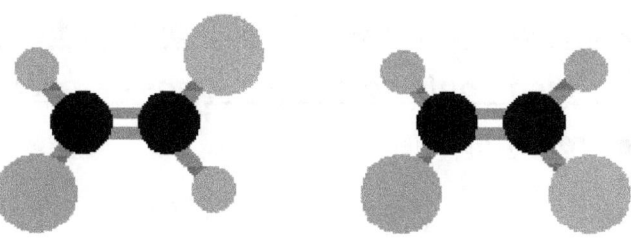

Here, the blue and green groups are either on the same side of the bond (cis isomer) or the opposite side (trans isomer).

Or you could go the whole hog and make everything different.

You still get geometric isomers, but by now the words cis and trans are meaningless. This is where the more sophisticated E-Z notation comes in.

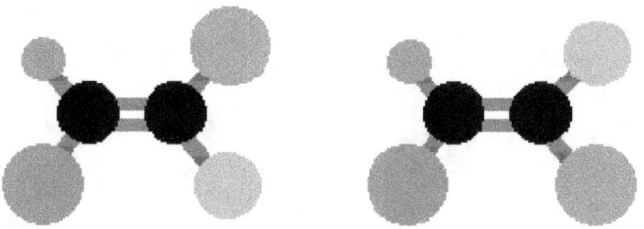

If there is no identical pair of groups attached to the same sp^2 carbon (geminal carbon) atom and If there is at least one pair of identical substituents

which are bonded to two adjacent sp^2 carbon atoms (vicinal), then only cis and trans nomenclature becomes relevant.

The E-Z system

The problem with the cis-trans system for naming geometric isomers

Consider a simple case of geometric isomerism which we've already discussed on the previous page.

trans-1,2-dichloroethene   cis-1,2-dichloroethene

You can tell which is the cis and which the trans form just by looking at them. All you really have to remember is that trans means "across" (as in transatlantic or transcontinental) and that cis is the opposite. It is a simple and visual way of telling the two isomers apart. So why do we need another system?

There are problems as compounds get more complicated. For example, could you name these two isomers using cis and trans?

Because everything attached to the carbon-carbon double bond is different, there aren't any obvious things which you can think of as being "cis" or "trans" to each other. The E-Z system gets around this problem completely - but unfortunately makes things slightly more difficult for the simple examples you usually meet in introductory courses.

We'll use the last two compounds as an example to explain how the system works.

You look at what is attached to each end of the double bond in turn, and give the two groups a "priority" according to CIP rules.

In the example above, at the left-hand end of the bond, it turns out that bromine has a higher priority than fluorine. And on the right-hand end, it turns out that chlorine has a higher priority than hydrogen

higher priority
on first C

higher priority
on second C

higher priority
on first C

Br   Cl

C=C

F   H

Br   H

C=C

F   Cl

higher priority
on second C

In the E-Z system the geometry is specified by the relative positions of the two higher priority substituents on the two carbons of the double bond. The priorities of the substituents are determined by using CIP rules. In the examples above, the molecules are divided into left and right sides. For each molecule, the group of higher priority on the left side double bonded carbon atom is first determined. Then you do the same thing for the right side double bonded carbon. If the two higher priority groups are on the same side of the double bond then the isomer is designated Z. If the two priority groups were on opposite sides of the double bond, the isomer would be designated E. The Z comes from the German word

Zusammen, meaning same, while the E comes from the German word *Entgegen*, meaning opposite.

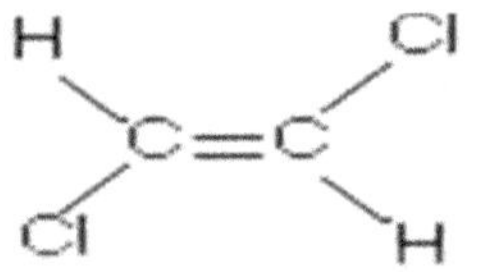

trans-1,2-dichloroethene

(E)-1,2-dichloroethene

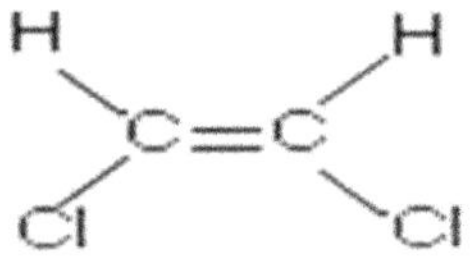

cis-1,2-dichloroethene

(Z)-1,2-dichloroethene

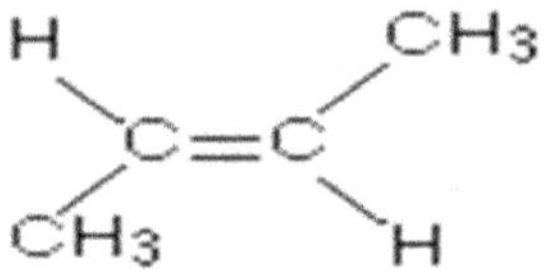

trans-but-2-ene

(E)-but-2-ene

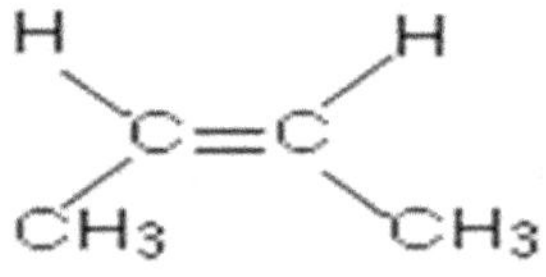

cis-but-2-ene

(Z)-but-2-ene

Summary

- (E)- : the higher priority groups are on opposite sides of the double bond.
- (Z)- : the higher priority groups are on the same side of the double bond.

CH₃ CH₃ / H H	1 CH₃ CH₃ 1 / 2 H H 2	1 CH₃ CH₃ 1 / 2 H H 2
Step 1 : split the alkene	**Step 2**: assign the relative priorities. The two attached atoms are C and H, so since the atomic numbers C > H then the -CH₃ group is higher priority.	**Step 3**: look at the relative positions of the higher priority groups: same side = Z, hence **(Z)-but-2-ene**.

Some problems

Q.1 The structure of (S)-2-fluorobutane is best represented by

(A) $CH_3\overset{F}{\underset{|}{C}}HCH_2CH_3$ (B) $H_3C\overset{F}{\underset{CH_2CH_3}{\overset{|}{C}}}-H$ (C) $H_3C\overset{H}{\underset{CH_2CH_3}{\overset{|}{C}}}-F$ (D) $F-\overset{CH_3}{\underset{CH_2CH_3}{|}}-H$

Q.2 The 2, 3-dichloropentane whose structure is shown is

$$\begin{array}{c}CH_3 \\ H-\!\!\!-Cl \\ Cl-\!\!\!-H \\ CH_2CH_3\end{array}$$

(A) 2R, 3R (B) 2R, 3S (C) 2S, 3R (D) 2S, 3S

Q.3 The S enantiomer of ibuprofen is responsible for its pain-relieving properties. Which one of the structures shown is (S)-ibuprofen ?

(A)

(B)

(C)

(D)

Q.4 Rank the following groups using the cahn-Ingold-Prelog system (4 is highest) :

$-CH(CH_3)_2$	$-CH_2CH_2Br$	$-CH_2Br$	$-C(CH_3)_3$
A	B	C	D

	4	3	2	1		4	3	2	1
(A)	C	B	D	A	(B)	A	D	B	C
(C)	C	D	A	B	(D)	C	D	B	A

Q.5 Give the configurations, respectively, of the following two molecules

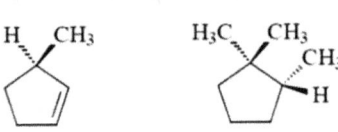

(A) R and R   (B) R and S   (C) S and S   (D) S and R

Q.6 Give the configurations, respectively, of the following two molecules

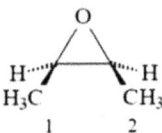

(A) R and R   (B) R and S   (C) S and S   (D) S and R

Q.7 Give the configurations of carbons 1 and 2, respectively, in the structure shown below.

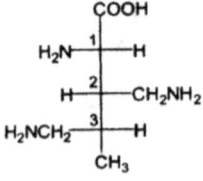

(A) R and R   (B) R and S   (C) S and S   (C) S and R

Q.8 Priority sequence of which alkyl group is maximum ?

(A) Tertiary   (B) Secondary   (C) Primary   (D) Methyl

Q.9 Assign stereochemistry to the following compound.

$$\begin{array}{c} COOH \\ H_2N \overset{1}{\underset{}{\rule{0pt}{1em}}} H \\ H \overset{2}{\underset{}{\rule{0pt}{1em}}} CH_2NH_2 \\ H_2NCH_2 \overset{3}{\underset{}{\rule{0pt}{1em}}} H \\ CH_3 \end{array}$$

(A) 1S, 2S, 3S   (B) 1S, 2S, 3R
(C) 1S, 2R, 3R   (D) 1R, 2S, 3S

Bibliography & References

1. *March, Jerry (1985), Advanced Organic Chemistry: Reactions, Mechanisms, and Structure (3rd ed.), New York: Wiley, ISBN 0-471-85472-7*

2. http://dictionary.reference.com/browse/stereo-

3. *Stephens TD, Bunde CJ, Fillmore BJ (June 2000). "Mechanism of action in thalidomide teratogenesis". Biochemical Pharmacology 59 (12): 1489–99. Doi: 10.1016/S0006-2952(99)00388-3.PMID 10799645.*

4. *Teo SK, Colburn WA, Tracewell WG, Kook KA, Stirling DI, Jaworsky MS, Scheffler MA, Thomas SD, Laskin OL (2004). "Clinical pharmacokinetics of thalidomide". Clin Pharmacokinet. 43 (5): 311–327. doi:10.2165/00003088-200443050-00004. PMID 15080764.*

5. *Francl, Michelle (2010). "Urban legends of chemistry". Nature Chemistry 2: 600–601. Bibcode:2010NatCh...2..600F. doi:10.1038/nchem.750.*

6. Anslyn, Eric V. and Dougherty, Dennis A. *Modern Physical Organic Chemistry*. University Science (July 15, 2005), 1083 pp. ISBN 1-891389-31-9

7. IUPAC, *Compendium of Chemical Terminology*, 2nd ed. (the "Gold Book") (1997). Online corrected version: (2006–) "gauche".

www.ingramcontent.com/pod-product-compliance
Lightning Source LLC
Chambersburg PA
CBHW060413190526
45169CB00002B/878